Stack Silver Buy Gold
For Beginners

How And Why To Invest In Physical
Precious Metals And Protect Your
Wealth When The Money Bubble Pops

By Jim Jackson

Disclaimer

This book is intended to be a general guide, to raise awareness, and to help people make informed decisions in the context of their own personal circumstance.

The author accepts no responsibility for any loss or injury be it personal or financial, as a result for the use or misuse of the information in this book. If you have any doubts or concerns after reading this book, please speak to a qualified person before taking any actions.

Contents

Introduction

Have you found yourself watching the news and becoming more than a little concerned with all the reports of failing economies? If you have, you are not alone.

There are plenty of experts and regular folks who are concerned with the direction of the current state of the economy in several major governments. A little research and it is easy to discover we are headed for a collapse and the current value of money is going to fall.

This book will explain why it is so important you start investing in silver and gold. Smart investors are looking to place their money in precious metals that hold their own value rather than put their money in a bank and hope it doesn't become worthless.

There are some things you need to know about buying silver and gold. Before you buy your first piece, read this book to learn all you need to know to begin stacking silver.

Chapter 1

Why Invest In Silver And Gold?

There has been a lot of talk lately about the state of the economies in some of the biggest governments all around the world. It is no secret the world's finances are on shaky ground. It leaves consumers on edge and looking for a way to protect themselves should the existing currency fall.

There is a quote by Thomas Jefferson that sums up the reason we should all invest in silver and gold.

"Paper is poverty...it is only the ghost of money, and not money itself."

Before we get into the many reasons a person should invest in either silver or gold, it is important to explain the difference between money and currency. They are not one and the same.

Currency: *a system of money in general use in a particular country*

Money: *a current medium of exchange in the form of coins and banknotes; coins and banknotes collectively*

Currency is the paper our founding father Thomas Jefferson was referring to. The dollar, euro, yen and so on, are all forms of currency.

Currency can lose its value and become worthless. It would only be worth the paper it was printed on. We know this is a strong possibility by looking back at the history of numerous currencies that have fallen.

A quick example of some of those fallen currencies are;

- German's papiermark that fell after WWI
- Zimbabwe dollar, before it's fall in the mid 2000s, 500 million Zimbabwe dollars was the equivalent of $2.50
- Escudo in Chile was replaced in the 80s after it became worthless after a massive rise in inflation

These are just a handful of examples of the more than 60 currencies that have come and gone over the past few centuries.

Money is a store of value, like silver or gold. These metals have real value, unlike currency which holds no actual value other than ink and paper. Currency can be printed every day, non stop. Silver and gold, or money, cannot simply be printed. There is an end to the supply of the metals.

Paper money is also referred to as fiat currency. Fiat currency is basically based on debt and is does not actually hold any value. If you were to follow the trail of the currency in your bank account, you would discover the money you earned at work has been borrowed from one bank and that bank borrowed it from another and so on.

History tells us that the currencies in place today will eventually fall.

Why?

Fiat currency is essentially worthless already. The only reason we are able to use it to buy the things we want and need is because everybody has faith that it is valuable. Once citizens and big banks start to lose that faith, the currency loses value. All the currency in the bank becomes worthless. BUT, the money, the gold and silver in your safety deposit box or stashed under your bed, is still valuable. It hold real value and not imaginary value.

Let's just briefly touch on why silver and gold are better investments than storing currency in a savings account.

- Gold has never lost its true value in the history of man
- Gold is an investment that you can keep to yourself—it doesn't go on the books anywhere and isn't taxed
- Gold and silver are currently extremely undervalued in today's world dominated by paper currency
- Gold and silver have more uses than just currency, increasing their desirability and value

Hedge Against Inflation

Skyrocketing inflation rates are typically to blame for the death of the currencies that have fallen over the past hundred years or so. Governments notice their buying power is weakening and the currency is becoming less valuable. In order to buy what is needed, more money is printed. Unfortunately, this just devalues the money even more. For example; a loaf of bread that costs a dollar today would cost a hundred dollars in an economy hit by hyperinflation. Silver and gold will not lose their value because of inflation. Gold and silver hold their own value and are not valued based on any government and that government's economy or political world.

Chapter 2

Gold VS Silver Which Is The Better Investment?

Once you have made the decision to invest in gold and silver, you are probably wondering which one is better. There are pros and cons to both. What it will really boil down to is personal preference, finances and availability.

Gold and silver buyers tend to take sides in this long-lived debate. Some are dedicated to one or the other, while others have recognized the wisdom of investing in gold and silver.

<u>Gold Buying Pros</u>

- Gold has steadily increased in value with no significant drops, only huge jumps
- Gold is considered more valuable simply because it has been a way of life for centuries
- There is more above-ground gold, which technically makes it easier to get your hands on
- People prefer gold jewellery, making it more valuable in a consumer's eyes
- A single gold coin is easier to carry than 20 silver coins, with the amounts equal in value

Silver Buying Pros

- Silver is much less expensive than gold, making it easier for the average person to buy
- Silver has many uses and is very easy to come by
- Silver is used industrially, which means it will appeal to a wider audience
- There is less known silver than gold, meaning there is an end in sight. With the demand for silver being higher than gold, it will become extremely valuable when silver mines are depleted
- It doesn't get old or decay, it is a solid metal that is going to be just as valuable, even more so, 20 years down the road

As you can see, both gold and silver have their good points. Gold has always been considered the most valuable precious metal, even though there is more of it than silver. Typically, a short supply makes something more valuable, but not in the case of silver and gold.

For the average consumer, silver is much easier to get their hands on. In fact, you can buy or even find old silver coins for a fraction of the price of their true value. Digging through grandma's old piggy bank is one way to do that. Coins made before the mid 1960s were made with real silver.

Historically, gold has been the metal of choice and only the wealthiest people had gold in their coffers. In an uncertain future, with gold still rather rare, silver will likely be just as valuable. Currency will mean nothing, but the amount of gold and silver you have stashed away will determine where you stand in the new world without currency.

Do not get hung up on buying in all one market. If you were investing in stocks, you would diversify your portfolio. The same method should be applied to buying silver and gold. If you can only afford a small of amount of gold, do what you can to buy that one bit and invest the rest of your available funds into silver.

One method many investors use is to buy low. Although gold prices don't drop too often, when they do, it is time to buy. Gold has proven to be a steady market that continues to increase. If you buy a gold piece today for $50.00, you are almost guaranteed it will increase in value on a yearly basis. Gold does tend to be more of a guaranteed investment, while silver tends to fluctuate more.

Gold is still the "gold standard" and many investors will tell the average investor to stick with silver. Gold buyers tend to hoard the gold they buy. It is a little tougher to come by in the resale market simply because the value continues to increase. Gold hoarders know that they should hold on to what they have, because it will only become more valuable.

Really, you are going to get more bang for your buck with silver. Silver will become more valuable. So, if you spend $10k on silver, it may one day be worth $100k. If you spend $10k on gold, which is really very little gold in today's market, it may only be worth $50k in the future. There isn't a lot of room for it to grow in value, while there is plenty of room for silver to increase in value.

Silver is due for a price explosion. It is a valuable metal that is running out. With so many different uses, it makes sense that it will only increase in value when the supply becomes critically low. In fact, here is a little tidbit of information that may help sway you.

Silver is the second most used commodity behind oil.

That certainly says something about how much the world relies on silver and just how valuable it truly is to our way of life.

Silver has many industrial uses that make it highly valuable and desirable. Some uses are as follows;
- Used in the pharmaceutical world for a variety of medicines
- Used as contacts in electronics
- Ground into a paste and used in auto manufacturing
- Used to make batteries
- Used to make jewellery
- Used in the photography world

Chapter3

Coins, Rounds Or Bullion The Pros and Cons Of Each

You are ready to buy a variety of silver and gold, but what kind of gold and silver should you buy. There are three main ways you can purchase your precious metal of choice. We will explore the pros and cons of each.

<u>Coins</u>

Most of the major economies mint coins that are meant to be collectible. The coins are made of gold or silver, but not as much as a bar or round is. You are paying a premium price for the collectible status and not so much the actual value of the precious metal. However, collectable coins are often easier to unload than a typical coin that is only worth  its weight—literally. Collectors will be willing to pay a higher price for a coin if it is deemed to be worth more because there are only a hundred produced. A collector may pay up to 80 percent more than the actual value of the metal in the coin.

Pros

- Coins are collectable and hold more value than the metal they contain
- They can also be used (but it would be silly to do so) at their face value. A quarter minted before 1965 will contain some silver and is more valuable than the .25 face value
- Coins hold less value and are therefore easier to trade when you need to buy small items
- You can buy junk coins in bulk and get a real bargain

Cons

- You would need to carry several bags of coins to buy large ticket items, trunks full if you are trying to buy something like a house or a car
- In today's market, coins are often valuable simply because they are collectible and not based on their actual metal content

Rounds

Rounds are a lesser known form of bullion. Instead of the familiar bricks you are used to seeing stacked in huge vaults in television movies, the gold is melted into round coins, but the rounds are not coins in a monetary sense. You would have a quarter or similar coin, but it would be made of almost solid gold or silver instead of

the government money that is used for currency today.

Pros
- Easy to stack and portable
- Low premium over spot price
- Manufactured by private mints
- Lightweight and can be used to purchase smaller items in a post SHTF, bartering world

Cons
- Not legal tender
- No collectible value
- Should be insured if you are storing a great deal
- Medium risk of buying counterfeit rounds; always buy from a reputable seller

Bullion/Bars
Gold bars are what you often see in Hollywood films. You see a bank vault stacked floor to ceiling with these bars and then you see daring thieves try to cart out as many of the heavy bars as possible. They don't typically get away with  many or have to use some pretty creative methods, like a forklift, to get away with a real fortune. Bars are sold in weights varying from a single gram to 10 ounces. So, 10 gold bars equals about 6 pounds. Not exactly easy to carry in your purse or wallet.

Pros

- Very easy to stack and store
- Bars tend to be favoured by serious investors
- Lowest premium over spot price
- Created by private mints—not government
- Cheaper overall because a producer only has to make 1 bar that equals 20 coins or rounds, best bang for your buck

Cons

- Not collectible
- Not legal tender in today's world
- Large bars would be difficult to barter with; they may be extremely valuable with no way to get "change" back
- Large and bulky, would need a dedicated space to store in an insured facility
- Increased risk of buying counterfeit bars

As you can see there are good and bad things about each. You will want to choose the option that fits your lifestyle and your need. If you are not comfortable putting several bars in a safety deposit box with your access being very limited, you may want to stick with the smaller rounds and coins. You will still need to secure your investment in a locking safe. Locking the safe to the ground is another necessary precaution.

Think about why you are investing and how you plan to use the gold and silver and make your decision from there.

Remember, you are not forced to pick one and stick with it. You can diversify and choose a little of all three.

Chapter 4

Buying Old American, Australian And British Silver Coins

You don't only have to buy coins and bullion from a dealer or printing press. You can buy old silver coins that may be sold by hobbyists, in pawn shops and other third party sellers. You may even be lucky enough to find these for a fraction of the price at yard sales and estate sales. A lot of people don't realize their old coins are actually worth money.

There are plenty of old American, British and Australian silver coins that are in circulation or being held in somebody's collection or old change jar. Old, collectable coins are given another name in the silver and gold buying world. They are referred to as numismatic coins. They are valuable because they are rare and coin collectors want them in their collection; not for the value of silver, but for the collectable value. Those you want to avoid.

Old American Coins

These old coins are often referred to as junk silver. Junk silver is the term used for old American coins that were manufactured before 1964. Despite being referred to as junk, these coins are anything but. No, they hold no collectable value, which makes them junk in a

collector's eyes. However, to the average silver stacker, they are worth the silver content, which is all that really matters.

While it is often rare to get change back from the supermarket or convenience store that contains any old coins put out before 1964, it does happen. When it does, consider yourself a very lucky individual because you just got a quarter back that is worth significantly more.

Junk silver American coins were made with 90 percent silver. The Jefferson nickel is an exception and is only made with 35 percent silver and the Kennedy half-dollar is only 40 percent silver. All in all, they still contain silver, which makes them important to the silver stacker.

Check out the list of coins that were minted before 1964 that you want to try and get your hands on. These coins are likely sitting in piggy banks, coin jars and even in the ground.

Quarters
- Washington-minted between 1934 and 1964, 90 percent silver
- Standing Liberty-minted between 1916 and 1930, 90 percent silver
- Liberty Head-minted between 1892 and 1916, 90 percent silver

Dimes
- Roosevelt-minted between 1946 and 1964, 90 percent silver
- Mercury (winged liberty head)-minted between 1916

and 1945, 90 percent silver
- Liberty Head-minted between 1892 and 1916, 90 percent silver

Half-Dollars
- Kennedy-minted between 1965 and 1970, 40 percent silver
- Kennedy-minted in 1964, 90 percent silver
- Franklin-minted between 1948 and 1963, 90 percent silver
- Walking Liberty-minted between 1916 and 1947, 90 percent silver
- Liberty Head-minted between 1892 and 1915, 90 percent silver

Silver Dollars
- Peace dollar-First minted in 1921 through 1928. Minted again in 1934 and 1935, 90 percent silver
- Morgan dollar-minted between 1878 and 1921, 90 percent silver

Buying junk silver is a bit of a crapshoot. You don't always want to buy it piece by piece. Buying bags of silver based on dollar amount is the most common way. So, you would buy $500 worth of quarters, but you are going to pay much more because you are buying the silver, not the face-value of the coins.

The bags are sold by weight. Let's assume you are going to buy $500 worth of coins at face-value. That bag of coins will weigh around 27 pounds. You will pay the silver price on that particular day or week. So, 27 pounds of silver coins is about 357 ounces of silver. An ounce of silver is typically about $20 an ounce. 357X20=$7,140. You are not buying coins, but silver. Expect to pay the price.

Australian Coins

Junk Australian coins are even more valuable than old American coins. The Florins, Shillings, Sixpences and Threepences issued between 1910 and 1945 were made with 92.5 percent silver.

Coins minted between 1946 and 1964 were made with 50 percent silver. With the introduction of decimalisation in 1964, the silver content was reduced to reflect the face value of the coin.

Australia and the rest of the world figured out sometime in the 1960s that more silver was being used than was being mined. It became apparent that using silver in coins was actually rather wasteful. The change from using silver in coins started early in the 1930s after the first World War and continued throughout the 1960s.

In 1966, Australia minted 50 cent coins that are now referred to as rounds. Those rounds were made with 80 percent silver.

A list of Australian coins that you will want to keep your eyes open for are as follows;

- 3 pence minted before 1947
- 6 pence minted before 1946
- Shilling minted before 1946
- Florin minted before 1946
- 50 cents, only made after 1966
- Crown, these were only produced in 1937 and 1938, they contain 92.5 percent silver

As with the old American coins, you can buy the junk Australian coins in bulk bags. You will pay for the weight of the silver and not the face-value of the coins.

Old British Coins

The UK also stopped using as much silver in their coins in the mid 1940s. This means, those boxes of coins in grandma's house that she has been holding on to for years are a lot more valuable than you may have originally thought.

Unfortunately, pre-decimal English coins that were minted in the 1950s and 1960s are essentially worthless. They are not legal tender and they are not made of silver. However, there are a few banks that will take them in and give you face value for the coins. From there, the coins are melted down and basically recycled.

Unlike Australia and the United States, the British cut back on the silver content in their coins much earlier. The coins minted before 1920 were made with 92.5 percent silver. Coins minted between 1920 and 1946 contain 50 percent silver. Don't completely ignore the 2nd generation coins. A little silver is still worth something.

You will find junk silver sales similar to the Australian and American coins. However, make sure you check the mint dates on the coins. Coins that were minted between 1920 and 1947 do not contain as much silver and therefore the weight and value needs to be adjusted.

The following is a list of coins you will want to add to your junk silver collection whenever possible.

- Threepence
- Sixpence
- Shilling
- Florin
- Half Crown
- Double Florins (1887-1890)
- Crown

Chapter 5

Where To Find The Best Deals On Gold And Silver

Gold and silver prices fluctuate almost daily. You will struggle to find a bargain or a really good deal on a commodity like gold and silver. While there are of course going to be surcharges and fees for going through a middle man, there are ways to avoid paying far more than you need to when buying your gold and silver.

It is important you know what you are buying and what it is truly worth. Do not walk into a coin shop without arming yourself with knowledge about that day's price of silver and gold.

While you will probably find silver and gold sellers online by the dozens, it is not a good idea to go that route. We will go through some of the best ways for you to buy silver and gold.

Online Auctions/Ebay
Ebay is one of those places silver stackers either really like or really hate. Many are quick to turn their nose up at sellers on eBay and assume it isn't a good deal. That is an unfair judgement and not entirely accurate.

Buying gold and silver pieces online through an auction site will likely increase your prices a bit. This is simply because you are adding an additional person between you and the manufacturer of the gold and silver pieces.

However, for many people around the world, this is really the only option. Without the ability to run downtown and visit a coin shop, this is one of the only ways they can buy silver and gold.

Obviously, it is important you research the seller and make sure they are legit. Check past feedback and the satisfaction of other buyers. Many mints, pawn shops and coins shops sell on eBay right out of their brick and mortar stores. These are excellent places to buy from and add a little more security to the transaction.

When you use Paypal or another credit card, you are also adding a layer of protection against fraud. You have options if you happen to buy from the one person who isn't selling real silver and gold. We will discuss how to determine whether you are getting the real stuff in the next chapter.

Ultimately, online places like eBay are not a bad option. You can find junk silver and gold pieces a lot easier than you would should you visit your local stores.

Coin Shops
Coin shops are an easy place to pick up silver and gold. You will find a variety of pieces that have already been weighed and are basically guaranteed to be exactly what they are being sold as. The variety of coins will range from junk coins all the way up to the collectable coins.

The coin shops are a middle man, so you will be paying a slightly higher premium. They buy the silver and gold from a mint and then they add on a fee to cover their overhead, which you will pay. It is important to know that even though you may check the price of silver on a given day, you will not get to pay that exact price. There is always going to be some kind of surcharge.

Developing a good relationship with a coin dealer is always a good idea. While it isn't fair to expect steep discounts, you may be able to develop enough rapport with the dealer to avoid additional fees that are passed on to other customers.

When browsing through the listings on eBay, there are some keywords you want to look for. A seller may be honestly telling you about the so-called silver, but what you are not able to understand is the silver being sold isn't silver at all. If you see one of the following words in a listing, avoid it.

- Replica or copy means the coin isn't silver, but is made to look like a silver coin
- Silver clad is another word for silver plated
- Silver plated is another metal with a silver coating
- 100 mills is misleading, but actually accurate, it is measuring the thickness of the silver plating on the coin, not the coin itself
- Nickel silver, sometimes referred to as German silver, isn't really silver at all, it is 60 percent copper, 20 percent nickel and 20 percent copper

Private Mints

There are several private mints that sell the rounds and bars. The mints have online shops, which makes it nice for everybody all around the world. However, some of the most popular mints, like JM Bullion, only ship within the United States. Some dealers will ship to Canada, but not all. There are typically mints available in every country.

Because you are ordering directly from the mint, you can expect to pay a bit less than you would from a local deal or other third party seller.

Private mints are not regulated by any government. They compete against each other, which is good news for the consumer. They work hard to keep their price over spot relatively low in order to attract customers.

It is important to read through the buying terms through a private mint. There are often minimum and maximum orders. You will also pay a premium shipping rate because of the insurance needed. This is another area of concern. If the order is lost, an insurance claim must be filed and that can take some time to recoup the loss, assuming the insurance was purchased high enough.

Classifieds

Checking the newspaper classifieds is another great way to find good deals on silver coins. You will find regular folks wanting to unload their silver without going through a dealer. This can save you money by cutting out a middle man. Look for junk silver sold in bulk. Always check the coins to determine whether they are truly worth what is being asked.

Craigslist
Many private sellers as well as brick and mortar coin shops will list their silver and gold pieces on Craigslist. This is an excellent way to find leads for what you are looking for. Always ask to see the coins in person and always meet in a public place. Make sure you take along your scale and magnifying glass before making a purchase.

Estate Sales
This is an excellent way to pick up some really great coins at very low prices. Often times, after a person has passed away, the family will sell all of their belongings. They may not realize what they have and sell you junk silver coins for face value or slightly above.

Check the classifieds and go early to any estate sales. You are not the only person looking to find a great deal on silver. While it is unlikely you will find gold, silver is always hiding in plain sight. You must be willing to dig around.

Coin Shows
Check with your local fairgrounds or other event centres. There are often yearly coin shows where collector's put some of their collectable coins up for sale as well their junk coins. Many of these sellers will be open for trades as well if you happen to have some coins you want to unload.

Many of these sellers will be willing to haggle a little on their prices. You need to go in knowing what the current price of silver is and what is considered junk coins and what are considered numismatic. Many dealers will want cash only as well.

Advertise in the Classifieds
There are plenty of people who have jars and bags full of old coins. They know the coins have some value but are not interested in silver stacking. Offer to pay a fair price for the silver and you may just find some willing sellers.

It is hard for silver stackers to understand, but there are plenty of people who are simply not interested in silver. Many people do not understand the need to hold silver and will get rid of it without a second thought. You may also find somebody who has fallen on hard times and needs to unload their silver to make ends meet.

Chapter 6

Checking The Purity Of Your Gold And Silver

Even though you may buy from a reputable dealer, it is always a good idea to check the purity of your gold and silver. Knowing how to do this is especially helpful if you are collecting junk gold and silver. If you are buying from a pawn shop or other dealer at an event, knowing how to test the product is essential.

Stamps/Markings

One of the quickest and easiest ways is to check the markings on the coin. It should be stamped with a series of numbers. For example; if there is a 975 on the coin or metal 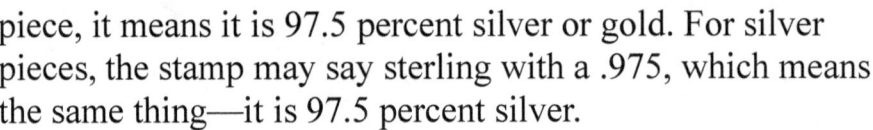 piece, it means it is 97.5 percent silver or gold. For silver pieces, the stamp may say sterling with a .975, which means the same thing—it is 97.5 percent silver.

The stamps can be rather difficult to see on small pieces and coins. It helps to have a magnifying glass to quickly find the stamp.

Electronic Testers

If you don't want to bother looking for stamps or maybe you have junk silver or gold that doesn't have a stamp, you can use an electronic tester. However, it is important you buy a tester that is tested and works. Don't buy a cheap tester from a questionable dealer. If your tester is not accurate, you could lose a great deal of money by selling off gold and silver that is actually much more valuable than your tester reported.

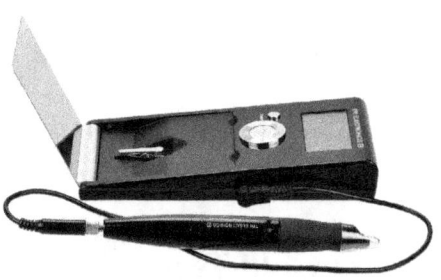

The electronic testers are very handy and are an excellent option if you have a great deal of metal you want to check.

Density Tests

This is a long, in-depth process that isn't all that appealing to the busy silver stacker. However, it is one way to verify your gold and silver content.

You will need a scale that measures weight down to the gram. Weigh each piece. Write down the number for each silver piece.

Next, fill a measuring cup or beaker that measures volume

with water. Drop the piece into the water. Measure how much the water moves up. The difference between the original water volume and the new water volume is the volume of the piece you put into the water.

Compare your two numbers. Silver has a density of 10.49 g·cm-3.

Visual Gold Test

Gold is a pretty stout metal and will not turn colours or rust. If it looks discoloured, it is likely gold plated and not actual gold.

You can also check the density of gold by giving it a float test. If it floats, it is not gold. If it floats even a little, it is not pure gold. Gold should always sink. This test isn't full proof. Other metals that do sink could be gold plated.

Your best option is to always buy from a reputable dealer. Use these tests as backups when you are not sure.

Acid Test

You can buy a special combination of nitric and muriatic acid to test silver coins. The acid should never be used to test numismatic coins as the acid will diminish the value. You will need to wear goggles and gloves when using the acid. Drop one to two drops on a coin you suspect may be fake. There will be a change in colour for real silver. If you suspect the coin is silver plated, use a file to rub off the outer layer and then test it.

Ping Test

Silver makes a unique, high pitched sound when pinged. Using a coin that you know to be silver, carefully tap it against the suspect coin. If you do not hear a distinctive ring sound, the suspect coin is likely fake. You will want to combine this test with another to make absolutely sure.

Magnets

This is a quick and simple way to check bullion. Hold a magnet over the coin. If the magnet sticks or is attracted to the coin, it isn't silver or gold. Iron or steel is often used in bars and rounds. If the magnet sticks, the bullion likely contains one of the metals.

Ice Test

Silver is an excellent heat conductor. Place an ice cube on top of the coin or bullion and wait to see what happens. If it is true silver, the ice should start melting immediately.

Bleach

Do not use this test on coins with numismatic value! Add a drop of bleach to the silver. If it starts to turn black, it is silver. However, a silver plated coin will also test positive as true silver. Combining this test along with weighing the coin is a good idea.

Conclusion

Buying silver and gold is one sure way to protect yourself should a major economy collapse. It is inevitable that fiat currency will become worthless. It could be in the UK or in the US, but if and when it does, the consequences will be far reaching. It will not effect just one corner of the world. It will have a ripple effect that will stretch out, all around the world to every single currency currently in use.

Silver stackers are aware of what is coming and are doing what they can to buy silver and gold to hedge against inflation.

You don't have to be wealthy in today's economy to start collecting silver and gold. Buying a little here and there ensures you will be one of the wealthy when fiat currency becomes worthless. All of those silver pieces will allow you to buy what you need. You won't be dependent on the money in your savings account. You won't be dependent on vouchers that are essentially worthless.

There is no real guarantee your money will be in the bank tomorrow. Whether it is FDIC insured or not, your money in today's world is standing on shaky ground.

Consider investing your money in gold and silver rather than putting it in stocks, bonds or as cash in a savings account. Gold and silver are truly the only items that hold their own value and are not dependent on any government's stability.

From The Author

Thank you for taking the time to read this book. As an author, I understand the importance of creating books which my readers will find both enjoyable and informative. If you have the time and feel generous, please don't hesitate to leave an honest review of this book..........Jim Jackson

Other Books By Jim Jackson

The Death Of Money

Surviving an economic collapse requires that you be prepared. This small guide will enable you to formulate a plan, allowing you to be proactive instead of reactive to a catastrophic financial crisis. In four chapters, you will gain invaluable knowledge and insight into what it takes to ensure you and your family have the tools necessary to survive the devastating impact of the loss of paper assets. Discover the skills you need to withstand the perils of a vulnerable financial system.

The Prepper's Grid Down Survival Guide

A major collapse of the power grid is inevitable. There are numerous scenarios that could cause a failed power grid that could leave large sections of the country or world in the dark. If you don't what could cause a massive power grid failure, you need to read the book. It isn't just the lights that go out. Everything will grind to a halt and it will be survival of the fittest, or in this case—the most prepared. Do you know what you need to prepare for a massive power failure that will put life as you know it in jeopardy? Can you feed your family with what you have in your house right now? Do you know what to do to take care of sanitation needs, water requirements and your comfort in general?

Don't be embarrassed if you don't have the first clue about what you would do if you were plunged into a blackout. Many people don't, which is why you need this book. It will guide you through everything you need to know to stay alive in the event of a major power grid failure. You will learn some valuable tips that will help you prepare for the imminent failure of the power grid. There is no time like the present to start preparing your home and your family to live and ultimately thrive a disastrous event like a failed power grid. Stocking up today, could save your life tomorrow.

www.ingramcontent.com/pod-product-compliance
Lightning Source LLC
Chambersburg PA
CBHW070743180526
45168CB00004B/1519

9 781516 957323